Sponsoring Editor: Anne Kelly
Developmental Editor: Rebecca Strehlow
Project Editor: Carol Zombo
Design Administrator: Jess Schaal
Text and Cover Design: Terri Ellerbach
Cover Illustration: Painting © Helen Hardin 1976, *Prayers of Harmonious Chorus.* Acrylic Painting. Photo © Cradoc Bagshaw 1994.
Photo Researcher: Carol Parden
Production Administrator: Randee Wire
Compositor: Interactive Composition Corporation
Printer and Binder: R.R. Donnelley & Sons Company
Cover Printer: R.R. Donnelley & Sons Company

For permission to use copyrighted material, grateful acknowledgment is made to the copyright holders on page 479, which are hereby made part of this copyright page.

Faces of Mathematics, Third Edition

Copyright © 1995 by HarperCollins College Publishers

All rights reserved. Printed in the United States of America. No part of this book may be used or reproduced in any manner whatsoever without written permission, except in the case of brief quotations embodied in critical articles and reviews. For information address HarperCollins College Publishers, 10 East 53rd Street, New York, NY 10022.

Library of Congress Cataloging-in-Publication Data

Roberts, A. Wayne (Arthur Wayne)
    Faces of mathematics / A. Wayne Roberts. -- 3rd ed.
        p.   cm.
    Includes index.
    ISBN 0-06-501069-8
    1. Mathematics. I. Title.
QA39.2.R6 1994
510--dc20   94-2399
                                                CIP

94 95 96 97 9 8 7 6 5 4 3 2 1

third edition
of Mathematics

A. Wayne Roberts
Macalester College

HarperCollins*College*Publishers

Contents

part *III* Reasoning and Modeling

part *IV* *Abstracting from the Familiar*

Preface

*I*t has long been held that anyone who aspires to be educated must study mathematics, and *Faces of Mathematics* is intended to be a source book for those who want to see what mathematics can contribute to a liberal education. In particular, it is addressed to college students who plan to take just one or two semesters of mathematics. Perhaps they want to satisfy a curriculum requirement, or perhaps they are prospective elementary school teachers who need a broad perspective on the field, together with some depth in understanding the underlying concepts of elementary mathematics.

A number of books have addressed this audience. Some try to survey the content of mathematics, offering a smorgasbord from which users may choose according to their tastes. Others emphasize the methodology rather than the content of mathematics. The former books are often intriguing but superficial; the latter are impressive but often too difficult for the audience we have in mind.

Those who believe, as I believe, that educated people should study mathematics know that mathematics can help us learn something about thinking itself: how to state our problems clearly, sort out the relevant from the irrelevant, argue coherently, and abstract some common properties from many individual situations. I wish to move toward these goals by steering a middle course among the available texts. Insofar as it is consistent with maintaining a light, readable style that will appeal to our audience, I have selected topics that can be presented in some depth. Moreover, I have continually addressed the larger contention that mathematics is the ideal arena in which to develop skill in the areas of organizing information, analyzing a problem, and presenting an argument.

A Word about the Title

The original title, *Faces of Mathematics*, was chosen for two reasons. The first was to emphasize the fact that mathematics was developed by human beings, real people with real faces. True, they may have had special talents, but on the whole they lived their lives subject to the same constraints as anyone else. Results in mathematics do not arise through divine revelation; they represent the hard work of individual men and women. The faces and brief biographies of many of the most significant contributors to this field appear in this book.

The second goal was to suggest that mathematics is like a finely cut diamond; it must be seen from several sides to be fully appreciated. Each view exposes a new face with its own distinctive features. Four of these faces—solving problems, finding order, reasoning and modeling, and abstracting from the familiar—reflect activities characteristic of mathematicians. The text is organized around these four faces.

An Emphasis on Involvement

A truism frequently seen in mathematics departments proclaims, "Mathematics is not a spectator sport," and I have tried in various ways to get readers involved in the stuff of the subject. To a large extent, I have done this by trying to engage the reader in solving problems.

So what's new about problems in a mathematics course? Plenty, because my purpose is not to develop particular computational skills, but to give insight into how one approaches problems. This enables me to replace the long lists of similar exercises designed to develop skills with puzzle type problems that serve my purposes just as well, and are much more engaging to work on. I believe, moreover, that the principles of problem solving I describe can be carried over into many of the areas of human activity where all of us are called upon to solve problems.

Part I of this book consists, therefore, of three chapters on problem solving. I seek to develop the heuristics that provide a framework for getting started, suggestions for methods that might be tried, and encouragement to think carefully about a proposed solution. These same themes are emphasized throughout the rest of the book, and reminders of problem-solving skills prompt students from the margins of most chapters. I also use a problem to open every section through the rest of the book, a problem chosen to be memorable (the kind that intrigues you, that you tell a friend about during lunch), as well as to be a kind of "hook" for wanting to read the material in that section.

Students should find in the lists at the end of each section in Part I a few problems that engage them in a sustained effort, that provoke discussion, and that require some careful written expression if their solution is to be understood. This kind of involvement is encouraged in Parts II, III, and IV of the book by including after the problem set in each section an outline of a project under the title *For Research and Discussion*.

An Emphasis on the Human Element

Consistent with some of the purposes described in the explanation of why this book is called *Faces*, the biographies of mathematicians scattered throughout the book focus not so much on their work as on other aspects of their lives. I have tried in my choices not so much to include the most important contributors (though that certainly was a consideration) as to emphasize that contributions to mathematics have been international in character, that interest in some problems has spanned generations, that women have made contributions and encountered resistance to their work, that the ugly specter of nationalism has affected the way the subject has grown, and so on. In short, I have tried to emphasize mathematics as human activity.

There are numerous things that most mathematicians would like their neighbors, friends, and even family members to understand about their subject. Many of them are explained in the essays that begin each chapter. Taken together, these fourteen essays give insight into how mathematicians perceive themselves and the work that they do.

Changes in the Third Edition

Those familiar with previous editions of this book might appreciate a summary of what is different in this edition.

The role of problem solving, always cited by users of the book, has been strengthened in three ways. First, Part I, the part that introduces problem solving heuristics, has been enlarged both in coverage and with new problems. It contains Chapter 1, "Getting Started," which focuses on what to do when you don't know what to do; Chapter 2, "Methods of Solution," intent on developing a mental checklist; and Chapter 3, "Reflecting on Solutions," reminding readers that there's more to do than check. Second, the ideas introduced in Part I are emphasized in exposition throughout the rest of the book. Finally, every section of the book now opens with a titled problem that I hope readers will find memorable in the ways explained above.

This edition reflects the author's agreement with that increasing number of mathematics teachers who have come to believe that students should be asked to do some research, writing, and discussion in mathematics just as they are in other classes. Thus, except for Part I where students have long lists of problems that provide for the activities just mentioned, every section of the book ends with a suggestion *For Research and Discussion*.

The essays mentioned above, found at the beginning of each chapter, are set forth with great hope that teachers will delight in assigning them, that students will greatly enjoy reading them, and that discussion of them will help students to better understand not only mathematics, but the people who work at mathematics.

The text has been brought up to date in a variety of ways. Biographic material now includes some living mathematicians; discussion of some recent advances (Fermat's last theorem, the four color problem) emphasizes that mathematics continues to grow as a discipline; outdated material (notably the chapter on computers) has been dropped; and the identification of problems where a calculator might be useful has been replaced with the idea that every student has and will make use of a calculator wherever that makes sense.

Advice to Teachers

This text can be used in a variety of ways. The book contains sufficient material for a full-year (two-semester) course. It is also easy to make selections for the typical semester course offered at many colleges. One-term courses that emphasize problem solving can be built around Chapters 1, 2, 3, 7, 8, and 12, for example. One-term courses that are more philosophical, with particular attention to clear thinking and precise writing, may use most of Chapters 1, 2, 3, 9, 10, 11, 12, 13, and 14. There are many other possibilities. The Dependence Chart that follows the preface will help you design a course to your liking. It illustrates the way chapters build on each other and cluster so that you may rearrange them in logical order.

Supplement Package

The new edition of *Faces of Mathematics* is accompanied by a supplements package intended to meet a range of teaching and learning needs.

Student Solutions Manual, prepared by the author, with complete, worked-out solutions to most odd-numbered section exercises.

Instructor's Manual has three components. In the first the author describes the way in which he handles classroom discussion and presentation of material on a section-by-section basis. This includes discussion-provoking questions and is helpful with difficult material that has been found to be effective. The second section includes two hour exams, each exam based on 100 points, exactly in a form that might be used in a class, for each chapter in the book. Finally there is a section giving answers to even-numbered exercises in the text.

HarperCollins Test Generator/Editor for Mathematics with Quizmaster is available in IBM and Macintosh versions and is fully networkable. The test generator enables instructors to select questions by objective, section, or chapter, or to use a ready-made test for each chapter. The editor enables instructors to edit any preexisting data or to easily create their own questions. The software is algorithm driven, allowing the instructor to regenerate constants while maintaining problem type, providing a nearly unlimited number of available test or quiz items. Instructors may generate tests in multiple-choice or open-response formats, scramble the order of questions while printing, and produce up to